Your Amazing
Itty Bitty™
Safety Book

15 Essential Steps for the Safe & Healthy Workplace Environment

Shannon,
You Are An inspiring
Leader! Best of
Health & Success.
Stay Safe & Healthy!

Stephen C. Carpenter, CSP

Published by Itty Bitty Publishing
A subsidiary of S & P Productions Inc.

Printed in the United States of America

Itty Bitty Publishing
311 Main Street, Suite E
El Segundo, CA 90245
(310) 640-8885

ISBN: 978-1-931191-42-5

I dedicate this to GOD Almighty.

I dedicate this to my fellow human beings
working day and night to provide for your
families. Stay safe and healthy!

I dedicate this to baby girl, little boy, momma
love, dad, mom, bro, sis, niece, my wonderful
family, amazing teachers & fabulous friends.
I love you all unconditionally.

One love

To get your **FREE GIFT** – the groundbreaking creative visualization – "Money-Mind Mastery: Make More Money With The Power Of Your Mind"– **Text** 55678. **Write the code**: "ITTY BITTY," in message box and enjoy.

Table of Contents
Steps Listed Alphanumerically

Introduction

"It has been observed at Occupational Safety and Health Administration (OSHA) Voluntary Protection Program (VPP) sites, and confirmed by independent research, that developing a strong safety culture has the single greatest impact on accident reduction of any process. A company with a strong safety culture typically experiences few at-risk behaviors; they also experience low accident rates, low turnover, low absenteeism, and high productivity." – OSHA

In this book you will find 15 essential steps listed alphanumerically that are necessary to create and sustain a safe and healthy workplace.

Safety and health resides within the larger frame work of occupational and environmental health and safety (EHS). The professional practice of safety and health incorporates elements of chemistry, biology, physics, law, epidemiology, physiology, geology, geography, toxicology, engineering, statistics, psychology, entomology, animal behavior, meteorology, and other disciplines.

EHS professionals are key stakeholders in the formulation of sustainable safety culture. They are strategic and tactical partners, laser focused on the timely identification and mitigation of EHS risks in the workplace environment.

Simple Steps

Step 1
360° Awareness

The concept of 360° Awareness, also referred to as Situational Awareness and Operational Awareness, pertains to the requirement for personnel to anticipate the unknown and know all aspects of the working environment at all times.

1. Conditions change constantly in the workplace, so personnel must be trained to anticipate, recognize, and respond to those changes on an ongoing basis.
2. Be ever-vigilant for the construction Fatal Four: Fall hazards, Electrical hazards, Struck-by hazards, and Caught-between hazards. They must be proactively mitigated!
3. 360° Awareness works to ensure our personal safety and the safety of our colleagues.
4. Workers around the world save one another's lives on a daily basis through their keen recognition of hazards and their proactive mitigation of hazards in the workplace.
5. When hazardous conditions are identified in the workplace, the hazards must be corrected within the immediate time frame.
6. At the end of the shift, hold a team huddle to assess the activities and chart the safest path.

Common 360° Awareness Phrases

- Always Ask Questions!
- Never Assume Anything!
- Stay Vigilant!
- Heads up!
- Be Aware!
- Know the Hazards!
- Know Your Emergency Egress Routes!
- Pay Attention at All Times!
- Do Not Go to Sleep at the Wheel!
- If You Do Not Know, You Do Not Go!
- Take 2 for Safety!
- YOU Have Stop Work Authority!
- See Something and Say Something!
- Take Safety Home with You!
- Stop! Look! Listen!

Most importantly: Hazards must be anticipated and recognized proactively on an ongoing basis. When hazards are identified, the hazards must be corrected immediately!

It is all about controlling energy and preventing the catastrophic release of potential and/or kinetic energy in myriad forms, including mechanical, chemical, hydraulic, pneumatic, radiant, thermal, electrical, magnetic, et al.

Timely corrective action is an important key to the safe & healthy workplace environment.

Step 2
Authentic Leadership

In an ideal scenario, every single person in an organization is a leader. That is precisely how EHS professionals want people to think. That is the way we want people to proceed – in the belief that they are indeed leaders in the workplace environment. It is not merely about your job title!

1. An effective leader proactively identifies and mitigates risk on a continual basis. This is a non-negotiable requirement.
2. All team members embrace 100% personal accountability and are 100% responsible for their decisions and actions.
3. When there is a victory, the leader gives it to the team. When there is a screw-up, the leader owns it and seeks resolution.
4. Leaders take full responsibility for everything within their span of control and even beyond. They do not point fingers. They own it!
5. The leader leads by example. He or she knows that people are paying more attention to what he or she does than what he or she says. That is a fact.
6. The leader does not berate his or her team members in a condescending fashion. The leader coaches and is open to being coached.

Leadership is a learned skill. As such, leaders continually hone their leadership skills in an ongoing effort to maximize their effectiveness.

- Leaders treat all personnel with whom they interact with dignity and respect.
- Leaders constantly hone their people skills and build personal relationships.
- Leaders are ideal strategic partners.
- Leaders are high achievers and tenacious.
- Leaders communicate effectively.
- Leaders build high performance teams.
- Leaders are essential to an organization.
- Leaders listen patiently & act decisively.
- Leaders know and practice forgiveness.
- Leaders have a positive mental attitude.
- Leaders create space for the flow of pure consciousness into the equation.
- Leaders always strive for excellence.
- Leaders transcend divide and rule politics to unite the team as a cohesive unit.
- Leaders generously serve others.
- Leaders continually coach, mentor, and teach those around them.
- Leaders model compassion and dignity.
- Leaders are laser-focused adding value.
- Leaders are willing to roll their sleeves up to work with the crew in the trenches.
- Leaders are firm, fair, and consistent.
- Leaders embrace continuous learning.
- Leaders read *Think and Grow Rich* and other personal development works.

Step 3
Effective Communication

In the world of safety and health, communication occurs when the Sender, the Receiver, and the Message are all on the same page.
1. A process called "double handshake" works to ensure effective communication.
   a. I share a message with you, and that is the first handshake.
   b. You repeat the message back to me conveying that the shared information landed, and that is the second handshake.
2. Effective communication keys win-win-win.
3. It is crucial that we communicate in a cool, calm, collected and professional manner. Once we raise our voices with one another and begin to argue, effective communication effectively ceases to exist.
4. High performance teams invest heavily in honing the communication skills and increasing the emotional intelligence quotients (EQ) of team members.
5. Strong people skills correlate with more dynamic interpersonal communication and catalyze those crucial conversations.
6. Effective interpersonal communication in the workplace is so fundamentally important that it makes or breaks the deal every single time!

Effective Communication

EHS professionals must continually network with one another in meaningful ways. LinkedIn®, the American Society of Safety Engineers (ASSE), American Industrial Hygiene Association (AIHA), American Conference of Governmental Industrial Hygienists (ACGIH), and the National Safety Council (NSC), serve this purpose well.

In the workplace, there is always an opportunity to reinforce the positive and address the negative in a constructive fashion. Communication can and does occur via a variety of means, including:

- Training
- Coaching
- Mentoring
- E-mails
- Phone conversations
- Conference calls
- Safety tailgate meetings
- Safety Committees
- In-person discussions
- Instant messaging dialogue
- Lectures with dialogue
- Signage and postings
- Safety bulletin boards; these represent an excellent forum for communication

Get the communication piece right and most everything else can flow from there!

Step 4
Emergency Preparedness

The importance of emergency preparedness cannot be overstated. It defines an essential part of the bigger picture for the survival of organizations as successful enterprises during and after adverse conditions and unplanned events.

1. All high performance companies recognize the importance of emergency preparedness and maintain evergreen written contingency plans to respond appropriately to the variety of potential emergencies.
2. Emergency preparedness incorporates business continuity, emergency response, emergency evacuation and business recovery.
3. Depending upon the actual situation, emergency preparedness may include pre-loaded mutual aid agreements with local, regional, state, federal, and supra-national authorities having jurisdiction (AHJ's).
4. Emergency preparedness also includes having a well-trained internal team with defined roles, responsibilities, and layered contingencies built into the structure. This is spelled out in the Emergency Action Plan.
5. Table top drills and field drills build muscle memory and good instincts. This ensures that personnel know their roles and are prepared to respond effectively in an actual event.

Potential emergencies include:
- Employee injuries and illnesses
- Offsite community impacts
- Hazardous materials (chemical) release
- Falling space debris
- Fires
- Explosions
- Building collapses
- Floods
- Tornadoes
- Landslides
- Earthquakes
- Hurricanes
- Tsunamis
- Typhoons
- Volcanic eruptions
- Armed robberies
- Terroristic activities
- Workplace violence
- Power outages and more

The alternative to emergency preparedness is unthinkable; an organization could be considered "safe" for years and literally have things turned completely upside down in an instant.

Emergency preparedness clearly represents a mission critical aspect of the safe and healthy workplace environment.

Step 5
Employee Training

Employee training pertains to the fundamental transfer of knowledge to employees regarding health and safety hazards associated with the work activities they perform. Effective safety and health training literally saves lives and forms the backbone of a dynamic EHS Program.

1. OSHA requires employee training to ensure that workers know the hazards in the workplace and know precisely how to protect themselves from exposure to those hazards.
2. EHS training occurs in conjunction with routine operations training, quality training, and other kinds of employee trainings.
3. For all workplaces there must be general EHS training and hazard-specific training. There are also levels of training that range from awareness level to competent person and qualified person level.
4. Team members must clearly comprehend the information and acquire performance-based knowledge required to ensure their safety and health in their workplace environment. EHS trainings are often bilingual or multi-lingual, including Spanish, Chinese, Vietnamese, et al
5. As new hazards are introduced into the workplace, new trainings must occur with updated information to capture any changes.

Examples of EHS Trainings for Employees

- Hazard Communication (GHS)
- Hearing Conservation
- Respiratory Protection
- Exposure Control Program / BBP
- First Aid / CPR / AED
- Electrical Safety
- Arc Flash / Arc Blast Prevention
- Radiation Safety (Non-Ionizing)
- Radiation Safety (Ionizing)
- Fall Prevention / Fall Protection
- Trenching and Excavation Safety
- Heat Illness Prevention Program
- Rigging and Signaling Qualification
- Ladder Safety
- Confined Space Entry
- Lock-out / Tag-out / Try-out (LOTO)
- Machine Guarding
- Defensive Driving
- Hazardous Waste Operations and Emergency Response (HAZWOPER)
- Stormwater Pollution Prevention Program
- New Employee Orientation / Onboarding
- Job Hazard Analysis (JHA / JSA)
- Daily Pre-Task Safety Plans (PTP / THP)
- Workplace Ergonomics
- Workplace Violence Prevention
- LASER Safety
- Asbestos and Lead Hazards
- Powered Industrial Trucks
- DOT Chemical Placarding Requirements

Step 6
Ergonomics Advantages

Human Factors aka Ergonomics pertains to the human-machine interface and defines the present and future of occupational safety and health.

1. Workers must report ergonomic-related discomfort early to ensure timely workplace evaluation and implementation of appropriate risk mitigation measures. **This is crucial!**
2. Ergonomics interventions cost money in the short-term; however, the long-term benefits and return on investment is tremendous.
3. Ergonomics programs have the unique positive potential to impact productivity, safety, quality, and morale simultaneously. This favorable outcome is a win-win-win!
4. Occupational Health Clinics are often on the front lines of effective ergonomics programs.
5. Ergonomics-related injuries include carpal tunnel syndrome, tenosynovitis, tendonitis, sprains, strains, and a variety of other injuries that affect muscles, tendons, ligaments, cartilage, and nerves in the human body.
6. A serious work-related back injury or carpal tunnel syndrome, for example, can result in permanent disability for the employee, devastating the worker and having profound adverse impacts on the rest of the team.

Ergonomics Advantages

Effective ergonomics risk mitigation focuses on reducing the incidence of soft tissue injuries in the workplace. Ergonomists proactively assess the workplace for risk factors and guide the implementation of risk mitigation measures.

Ergonomics risk factors include:

- Repetitive motion (frequency and duration of tasks / work-rest regimen)
- Awkward postures
- Static postures
- Excessive reach
- Excessive force
- Vibrations
- Lighting conditions
- Masses of objects being manipulated
- Contact stress, such as sharp edges
- Temperature extremes
- Interference with circadian rhythms
- Free time activities (hobbies / sports)

Serious ergonomics injuries can be life changing and extremely debilitating for affected personnel.

The focus of Occupational Health Clinics on the overall health, safety, and wellness of employees aligns perfectly with ergonomics. This provides for truly brilliant synergy in the workplace.

A robust Ergonomics process represents an integral aspect of any serious EHS program.

Step 7
Hierarchy of Controls

The Hierarchy of Controls comprises a roadmap to EHS risk mitigation. This speaks to the actual methodologies employed to correct hazardous conditions and prevent occupational exposures.

1. Wherever possible, the EHS professional wants to substitute or engineer exposures out to eliminate hazardous conditions altogether.
2. Whatever the situation, **engineering controls** often work in a successful partnership with **administrative controls**, and **personal protective equipment (PPE)** to ensure that all potential exposures are mitigated.
3. Administrative controls play a vital role that essentially brokers the deal in the field for the successful implementation of any effective safety and health program.
4. PPE, aka the "last line of defense," forms a crucial barrier to protect the workers from safety and health hazards in the workplace.
5. On construction sites, for example, minimum PPE requirements oftentimes include:
    a. Sturdy work boots
    b. Class II high visibility vest
    c. Safety glasses
    d. Hard hat
    e. Hearing protection devices

Examples of Engineering Controls include:

- Install machine guards to mitigate exposure to moving parts and pinch points
- Substitute a less hazardous chemical
- Install steel plate shoring in excavations
- Isolate a source of noise with a robust sound attenuation enclosure
- Install engineered railing systems to mitigate potential fall hazards
- Install a local exhaust ventilation system
- Implement adjustable workstation designs to accommodate a wide range of body sizes
- Implement engineered Sharps devices to prevent exposure to blood-borne pathogens
- Isolate biohazards in biosafety cabinets
- Install electrical interlocks to isolate process
- Install state of the art fire suppression system

Examples of Administrative Controls include:

- Worker safety and health training
- Signage and postings
- Labeling of hazardous materials
- Red danger tape / yellow caution tape
- High visibility cones / candlesticks
- Standard Operating Procedures (SOP's)
- Job Hazard Analyses (JHA's)
- Daily Pre-Task Safety Plans (PTP's)
- Methods of Procedure (MOP's)
- Policies & Procedures
- Injury & Illness Prevention Program (IIPP)

Step 8
Housekeeping

Housekeeping represents a fundamental premise of the safe and healthy workplace environment. The most organized workplace correlates with the most productive and safest workplace. Take care of your people, and they will take care of you!

1. Housekeeping can be used as the tip of the spear to build and sustain a positive and proactive safety culture in the workplace.
2. Slips, trips, and falls from poor housekeeping are a primary source of workplace injuries.
3. Excellent housekeeping not only protects the safety and health of workers, it improves quality, production and morale on the job.
4. Excellent housekeeping manifests in a variety of forms in the industrial environment. Lean manufacturing concepts borrowed from Japan articulate a path forward called 5S in English.
5. Introducing 5S: Sort, Set in order, Shine, Standardize, and Sustain.
6. 5S comprises an auditable protocol and a valuable tool in the worthy battle of bringing order to chaos in the workplace environment.
7. Marry 5S and Kaizen to your robust Rewards & Recognition Program and watch authentic measures of safety, quality, productivity, and success skyrocket!

Housekeeping and 5S
This pertains to lean manufacturing methodology
and the Oliver Wight "Class A" state of mind.
The 5S's are couched within the context of the
continuous quality improvement loop consisting
of a PDCA framework:

- Plan
- Do
- Check
- Act

*The vital importance and supreme value of good
housekeeping must not be underestimated in the
occupational setting.*

A Union Journeyman Carpenter I met years ago
lost a very close friend on the job due to poor
housekeeping. His friend stepped through an
unguarded floor opening that was unrecognizable
due to clutter, insufficient lighting, and poor
housekeeping conditions in the work area.

"Housekeeping! Housekeeping! Housekeeping!"

"See it! Own it! Do it!"

"See something! Say something!"

**If I see something, then I am obligated to do
something. I cannot merely ignore the hazard.
I might save my colleague's life by speaking
up. In the future she / he might save my life.**

Step 9
Incident Investigation

Incident investigation is the follow-up required after an incident occurs on the job. Incidents pertain to all personal injuries, property damage, environmental impacts, and any other unwanted or unplanned event, including "near miss" events.

1. As in any emergency situation, the formal incident investigation takes place after the emergency has been cleared.
2. In the construction safety world, companies often speak of workplace environments free of injuries, impacts, and incidents. Three fundamental tenets define this powerful belief system:
   a. All incidents, injuries and impacts are preventable.
   b. No degree of incident, impact or injury is acceptable.
   c. Everybody goes home safe and healthy at the end of his / her shift.
3. These relatively self-explanatory concepts only work in a blame-free environment.
4. The Root Cause Analysis (RCA) process attempts to reveal the contributing causes and a specific root cause of the incident.
5. The RCA process represents a fact finding mission and not a fault finding mission!

Whenever an incident, impact or injury gets investigated by the team, it must always be a fact finding mission and not a fault finding mission.

Incident investigations parallel investigations performed by the police, OSI, FAA or FBI.

- Evidence gets collected
- Interviews take place
- Photographs are taken
- Samples are analyzed
- Documents are reviewed
- Video and audio feeds get reviewed
- Rigorous detailed discussions occur oftentimes in the form of an RCA and any follow-up meetings

Ascertaining both root causes and contributing causes, the RCA also identifies action items for implementation within prescribed time frames to prevent the re-occurrence of such incidents.

The gold that can be mined from the incident investigation process pertains to the concept of "lessons learned" derived from the investigation results.

"Lessons learned" can be shared to prevent similar future outcomes, ultimately benefitting stakeholders throughout the organization and even across the industry.

Step 10
Industrial Hygiene (IH)

Industrial hygiene comprises the art and science of the anticipation, recognition, evaluation, and control of health hazards within and emanating from the workplace environment.

1. Industrial hygienists (IH's) focus on the mitigation of exposure to hazardous materials and prevention of *xenobiotics* from entering the human body via the 4 routes of exposure.
2. IH's implement strategies and methodologies that leverage the Hierarchy of Controls to prevent occupational exposure, community exposure, and environmental exposure.
3. IH's also focus on mitigating environmental concerns such as chemical hazards, biological hazards, radiological hazards, hazardous wastes, universal wastes and medical wastes.
4. Industrial hygienists protect humanity from blood-borne pathogens such as Hepatitis B, Hepatitis C, HIV, Bird Flu, Ebola, and a host of other bad actor viruses and biologicals.
5. The professional certification for industrial hygienists is Certified Industrial Hygienist (CIH); the analogous certification for safety is Certified Safety Professional (CSP).
6. IH's are extremely dedicated to creating a verifiably safe and healthy workplace.

There are four (4) Routes of Exposure:

- Inhalation
- Ingestion
- Absorption
- Injection

Using the Hierarchy of Controls, the Routes of Exposure must be guarded against exposure to a host of health hazards, including the following:

- Chemicals
- Carcinogens
- Teratogens
- Hazardous noise
- Non-ionizing radiation
- Ionizing radiation
- Ergonomics hazards
- Blood-borne pathogens (BBP)
- Viruses
- Molds
- Bacteria
- LASERS
- MASERS
- Zoonotic diseases
- Nanoparticles

These health hazards manifest in myriad forms of dusts, mists, fumes, vapors, fibers, gases, liquids, waves, particles, biologicals, energies and more.

Industrial hygienists are truly the most unsung heroes of the occupational setting!

Step 11
Management Systems

EHS management systems represent auditable programs that enable organizations to track improvements proactively and focus resources on risk mitigation in a more strategic fashion.

1. The management system also contains core aspects to guarantee sustainability is built into the functionality of the process.
2. EHS management systems can be a strong ally for EHS professionals partnering with management in the ongoing endeavor to improve performance and reduce risk in a methodical and well-organized manner.
3. EHS Management systems consist of several key elements required for detailed evaluation to benchmark EHS performance and optimize internal risk assessment activities.
4. EHS Management systems reside within the continuous quality improvement framework.
5. Management systems provide an organization with a risk-based means of identifying status today while charting a clear path forward of measureable improvements for 3-5 years.
6. "Compliance management" comprises a most critical and challenging system element.
7. Integrated and automated data management systems are essential to facilitate the process.

An effectively implemented EHS management system provides:

- A well-defined EHS management system intent is for an organization to go from bad to good, good to excellent, and excellent to outstanding, within a prescribed time frame.
- The system provides a means to obtain repeatable improvements and measurable results relative to authentic risk mitigation in the occupational setting.
- The OHSAS 18001 (Occupational Health and Safety Assessment Series) certification system is administered by the ISO (International Organization for Standardization) based in Geneva, Switzerland. The equivalent definitive Environmental Management System is ISO 14000, and ISO 9000 pertains to the Quality Management System.
- The integration of technology, cutting-edge software, and tools for assessing, measuring and improving occupational safety and health is highly recommended!
- **Embrace the technology** and implement the tools accordingly as you continually improve safety and health according to the parameters articulated in the EHS management system.

Step 12
Permits, Licenses, and Administration

There is an age-old saying in safety, "What gets measured gets done." It is one of the reasons that EHS professionals insist upon there being a paper trail or electronic trail for so much of the work in the world of safety and health.

1. Professional qualifications are necessary to perform specific tasks. "Competent Person" and "Qualified Person" are important designations that have specific roles and responsibilities attached to them.
2. Make sure you have the right person with the right knowledge for the right job.
3. Subject matter competency is attained by workers through hazard-specific trainings and on the job experience regarding a particular risk profile or hazard classification.
4. Training also comes with experience in the form of on-the-job training where the value of apprenticeships in the trades is manifest.
5. The administrative aspects of EHS programs are arguably as important as the other components, including hazard recognition and corrective action. Regulatory agencies mandate rigorous recordkeeping. As for EPA and OSHA, if there is no written record, then it is as if the events never occurred.

Permits, Licenses, and Administration

OSHA permits are required to perform a variety of activities in the workplace, including asbestos removal, demolition, excavations, pressure vessel operation, construction and elevator permits, tower crane erection and operating permits, et al.

Specific permits for organizations may include confined space entry, hot work, energized electrical work, fire life safety (FLS) impairment, excavation, hazardous waste permits, et al.

The Competent Person designation specifically applies to personnel in fall protection, trenching and excavation, and scaffolding inspection. The Qualified Person designation applies to Rigging, Scaffold Erection and other high risk activities.

Licensing requirements exist for mission critical jobs like operators of cranes, powered industrial trucks, manual elevated work platforms, and powder actuated tools. The safe performance of these dangerous jobs mandates State licensure.

Workers Compensation and additional insurance requirements necessitate diligent recordkeeping. Medical surveillance records need to be accurate and up-to-date. The records must be kept for 30 years beyond the employment of the workers. In light of the latency periods for certain hazardous chemicals, this is not only the law but a moral obligation to the workers and their families.

Step 13
Routine EHS Inspections

Inspections represent an extremely important tool for the identification of hazardous conditions and unsafe behaviors in the workplace environment.

1. The key to a successful EHS inspection program is that once hazards get identified, the hazards get corrected in a timely manner.
2. There are different kinds of inspections, for example: localized inspections of specific pieces of equipment, comprehensive wall-to-wall inspections, PPE inspections, and audits.
3. Correcting hazards usually means a capital investment. The short-term safety and health capital expenditure yields exponential long-term safety, health, quality, production, and morale returns. **This represents true value!**
4. Once the affected workers see that the EHS program is not smoke and mirrors, but rather an authentic, value-added means to mitigate hazards to make the workplace safer, the workers will take ownership of the program as strategic and tactical partners. This is a lynchpin for your successful safety program.
5. The most effective EHS inspections involve a cross-functional team including stakeholders from top management, middle management, frontline management, and the employees actually working in the trenches.

Routine EHS Inspections

Involve all levels of the chain of command in the inspection process to cultivate team camaraderie and hone stakeholder hazard recognition skills. Team up and manage the EHS risks proactively!

When you have front-line ownership of the EHS inspection process by the managers and workers, combined with unyielding management support, **you have the best recipe for a safe and healthy workplace** and a high performance organization.

Everything must be inspected prior to use:
- Equipment
- Machinery
- Power tools and power cords
- Vehicles
- Personal Protective Equipment (PPE)
- Harnesses, lanyards, Yo-Yo's, et al.
- Facilities
- Shop floors
- Factories
- Laboratories
- Offices, and more

To obtain their maximum benefit, inspections should occur in both a planned and unplanned manner. EHS inspection processes correlate with proactive risk identification and mitigation. As such, they can serve a powerful cross-functional purpose of making the desired routine a routine.

Step 14
Safety Culture and Morale

Safety culture is created and developed over time through dedicated commitment, follow-through, and proof of concept. It equates with meaningful relationships, established trust, and an authentic investment in employee well-being.

1. Treat every person in the organization with the utmost dignity and respect, and you are on the way to a high-performance culture.
2. **Morale is the secret ingredient** that defines true greatness in the corporate community. Morale can literally move mountains in an environmentally friendly fashion!
3. The bottom line is that the trained workers in the trenches know the hazards, and are true subject matter experts. Acknowledge them and heed their recommendations!
4. Engage your employees in the safety process, set them up for success, and empower them to model and share that powerful knowledge on an ongoing basis. Involve the team with origination of concept to bolster ownership.
5. Safety culture always distills down to people and interpersonal relationships. If you are choosing to make health and safety personal, then you are choosing to be successful!
6. Outstanding safety performance corresponds with outstanding business performance.

Safety Culture and Morale

- While all incidents, impacts, and injuries are preventable, achieving and sustaining the goal of a safe and healthy workplace environment is not an easy task.
- Management commitment, plus worker engagement, equals successful outcome.
- Make safety an **organizational value** and not merely a priority.
- There must be authentic investment, including financial, by management, in the safety and health program in order to foster and sustain positive safety culture.
- All stakeholders' contributions should be recognized, considered, and included in the safety and health process.
- Create an emotional connection to safety!
- Make the whole thing personal with the team members and model the successful attitude and performance you expect.
- Problems arise when workers start to cut corners in order to meet artificial time frames or begin to "fall asleep at the wheel" due to a lack of engagement.
- Workers are people and your most valuable asset. Employees are not merely human resources, but rather sentient human beings with families and friends who care about them & their well-being.

Human beings tend to be loyal. When you take excellent care of your people, your people will take excellent care of you. It's a Respect Game!

Step 15
Safety Integration

Safety integration engages the team in all aspects of how we do business and defines the optimal recipe for success. That means safety not only comes 1st, but safety comes 1st, 2nd, 3rd, 4th, 5th, 6th, 7th, 8th, and at every single junction of our decision-making processes, and accompanying actions in the workplace.

1. Safety is built into the muscle memory and consciousness of the way we do business.
2. Safety comprises a non-negotiable aspect of the routine. Focus your strategic vision on your people. Train them and set the team up for success. Encourage them and support their leadership and participation relentlessly.
3. It is always ALL about the team! Team is an acronym for the awesome phrase, "Together Everyone Achieves More!"
4. Know your hazards and establish well-funded cross-functional safety and health committees to mitigate those hazards, communicate, and facilitate the vital flow of information.
5. Implement an exciting and fair Rewards and Recognition Program to seal the deal.
6. EHS professionals must take every leadership opportunity available to them and choose to be brave when others choose to be scared.

29

Safety Integration

Safety integration means EHS requirements
reside in the bull's-eye of the target for why we
do business and how we do business.

- The team works together as one to ensure
  we all go home safe, healthy and alive to
  our wonderful families and friends.
- Everyone is fundamentally responsible
  and accountable to the safety process.
- The team owns safety on the floor and
  seamlessly includes EHS considerations.
- Personnel integrate emotional intelligence
  (EQ) with the intelligence quotient (IQ).
- A sustained proactive approach to EHS
  risk mitigation drives engagement and
  ensures ownership by stakeholders.
- As leaders, EHS professionals play a
  central role in the integration of safety &
  health into the culture of a company.
- Say what you mean; mean what you say!

In the ideal world, EHS requirements reside right
up there on the main stage with production and
quality, in the bright lights. The marriage of
these key elements defines a powerful sweet spot
in the high performance workplace environment.

The fundamental reality must always prevail …

Every employee goes home safe and healthy 100% after every shift!

You've finished. Before you go…

Tweet / share that you finished this book.

Please Five "Star Rate" this book on Amazon, if you found it of value.

Reviews are solid gold to writers. Please take a few minutes to give us some itty bitty feedback on this book.

32

ABOUT THE AUTHOR

Stephen Carpenter obtained his Master's degree in the Environmental Health Sciences from the UC Berkeley School of Public Health in the 1990s. Since that time he has worked as an EHS professional around the world at organizations within oil & gas, government, transportation, biotechnology, astronomy, and the high tech industry.

He has been fortunate to work with several outstanding EHS professionals along the way who have shown him the ropes & graciously guided his professional development. These people include his truly amazing instructors at UC Berkeley, who initially inspired him to go into the safety and health profession.

Stephen presently works on one of the most incredible construction projects in the United States of America. He considers it an honor and a privilege to be part of the dynamic project execution team. He is grateful for the opportunity. It is ALL about relationships!

Stephen believes in the fundamental right to a safe and healthy workplace environment for all employees. This is truly a family matter. He makes it his personal business to drive EHS requirements in a professional manner.

Other Amazing Itty Bitty Books

- **Your Amazing Itty Bitty™ Travel Planning Book** – Rosemary Workman
- **Your Amazing Itty Bitty™ Cruise Diary** – Itty Bitty Books
- **Your Amazing Itty™ Bitty Weight Loss Book** – Suzy Prudden and Joan-Meijer-Hirschland
- **Your Amazing Itty Bitty™ Food & Exercise Log** – Suzy Prudden and Joan Meijer-Hirschland
- **Your Amazing Itty Bitty™ Astrology Book** – Carol Pilkington
- **Your Amazing Itty Bitty™ Little Black Book of Sales** – Anthony Comacho

Coming Soon

- **Your Amazing Itty Bitty™ Tax Audit Prevention Book** – Nellie T. Williams, EA
- **Your Amazing Itty Bitty™ Business Tax Book** – Deborah A. Morgan, CPA
- **Your Amazing Itty Bitty ™ Book of QuickBooks® ShortCuts** – Barbara Starley, CPA
- **Your Amazing Itty Bitty™ Heal Your Body Book** – Patricia Garza Pinto
- **Your Amazing Itty Bitty™ Marijuana Manual** – Kat Bohnsack

35

With many more Amazing Itty Bitty Books
to come …

Stephen
415.513.8025
AmazingSafety.toy

24089665R00028

Made in the USA
San Bernardino, CA
11 September 2015